The Greatest
Guessing Game

A Book About Dividing

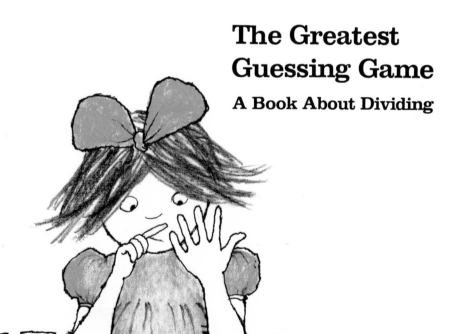

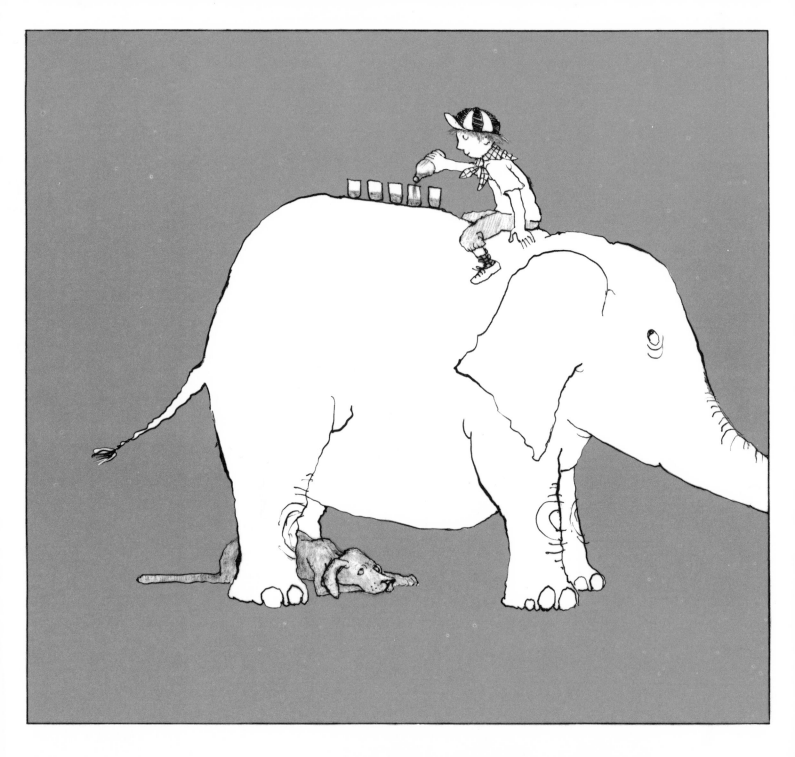

The Greatest Guessing Game

A Book About Dividing

By Robert Froman   illustrated by Gioia Fiammenghi

Thomas Y. Crowell   New York

YOUNG MATH BOOKS

Edited by Dr. Max Beberman, Director of the Committee on School Mathematics Projects, University of Illinois

Edited by Dorothy Bloomfield, Mathematics Specialist, Bank Street College of Education

Library of Congress Cataloging in Publication Data. Froman, Robert. The greatest guessing game. SUMMARY: Explains the process of division by comparing division to a guessing game. 1. Division—Juv. lit. [1. Division] I. Fiammenghi, Gioia. II. Title. QA115.F76   513'.2   77-5463   ISBN 0-690-01376-0   1 2 3 4 5 6 7 8 9 10

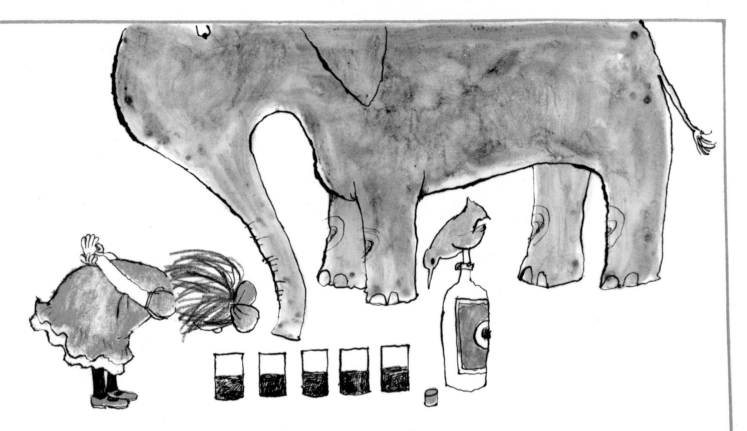

The Greatest Guessing Game

A Book About Dividing

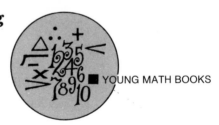

YOUNG MATH BOOKS

There are many different kinds of guessing games.
Someone sneaks up behind you and puts both hands over your eyes. You are supposed to Guess Who.

Or a friend ties a blindfold
over your eyes, turns you
around three times, then
leads you off several steps
and stops you. This time you
are supposed to Guess Where
you are.

There is another guessing game you sometimes play without even knowing that you are playing it. It is a game you will be playing all your life.

Suppose you have a bottle of root beer and two friends you want to share it with. You want each to have the same amount as you.

You get three glasses and pour root beer into the first one until it is about half full. Then you pour from the bottle into the other glasses.

If there is more than enough to fill all three glasses halfway, you go back and add to each glass a little at a time.

If there is not enough to fill all three glasses halfway, you pour from one glass into another until all are about even.

You hardly notice that you are guessing when you do this kind of dividing. You just estimate how much root beer there is for each of you, then change your estimate if it turns out to be wrong.

You sometimes play the same kind of guessing game when you are working with numbers. Suppose that instead of a bottle of root beer you have twelve walnuts. You want to share them with those same two friends. And you want each of the three of you to have the same number of walnuts.

One way to decide how many there are for each is by guessing. You don't have to guess right the first time. You can have as many guesses as it takes to find out what you want to know.

You might guess first that each of you can have five walnuts.

You count out five in one pile.

Then you count out five in a second pile.

This leaves only two for the third pile.

Now you know that the guess is wrong. So you guess again.

You know you did not
have enough for five each.
So you try a smaller number.
This time you might guess that
each of you can have three walnuts.
You count out three in one pile,
three in a second pile, and three
in a third pile. You have three left over.

You probably would not guess
again after this. You would just put
one walnut from the three left over
in each of the three piles. And you
would find out that there are enough
walnuts for each of you to have four.

That was a pretty easy one. You might have guessed four the first time. If you happened to remember that three times four makes twelve, you would have known right away that there were enough walnuts for four each.

But when you turn division into a guessing game, you do not have to remember the multiplication table. And you do not have to guess right the first time.

Suppose that the man next door offers you six dollars to clean up his yard. You get three friends to help you do it. You agree to divide the money so that each of you gets the same amount.

You do the work. The man gives you six dollar bills. How will you divide them?

You would probably know that you'd each get at least one dollar. But when you count out four dollars—one for each of you—there are two dollars left over.

What can you do with them?

One thing you can do is take them to a store and have them changed into half-dollars. For each dollar you would get two half-dollars. For the two dollars you would get four half-dollars.

You would not even have to guess how to divide the four half-dollars. There would be one for each. So each of you would get one dollar and one half-dollar—or $1.50.

Then suppose you and your friends
decide to go on and try to earn some
more money together. Another friend
joins you. That makes five of you.
You find another yard to clean up.
You also collect a lot of aluminum cans
to take to a place that pays for them.

And you help a woman wash and wax and polish her car.

15

When all this is done, you find that the five of you have earned a total of $8.27. How would you divide it into five equal shares?

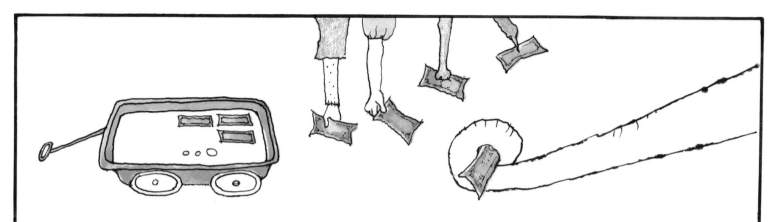

Since there are five of you, and more than five dollars, you would probably know that there is enough for more than one dollar each. You could take eight dollar bills and count out five piles with one bill in each pile. You would have three dollars and twenty-seven cents left over.

If you have these three dollars changed into half-dollars, you would get six half-dollars. It is easy to see that this is enough for one half-dollar each. And there is one left over.

What about this extra half-dollar? Can you think of any easy way to divide it among the five of you?

The easiest way would be to change it into dimes. A half-dollar is worth five dimes, so you would have one dime apiece.

That leaves only the twenty-seven cents. Can you think of a way to divide this amount?

You can think of it as a quarter and two pennies. A quarter is worth five nickels. So the quarter can be divided into one nickel for each of the five of you.

Now each of you has one dollar, one half-dollar, one dime, and one nickel. That adds up to $1.65. And there are two pennies left over.

Leftovers like these in division are called "remainders." And remainders can be difficult to deal with.

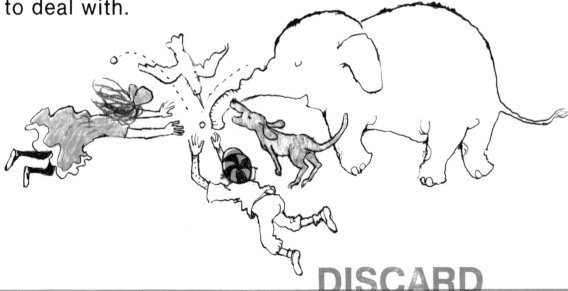

But in this guessing game the remainder is no trouble at all.

You could buy a two-cent piece of candy with the pennies and divide it five ways.

Or you could draw straws, and those who draw the two longest straws would each get one penny.

Or you could give the pennies away. See if you can think of other things you might do with them.

In this guessing game you work on the important parts of what you are dividing. When you are through dividing the important parts equally, you can do whatever you wish with the rest.

Sometimes it is good to have a remainder that you do not divide up.

Suppose that you have a club with some friends. There are eleven members altogether. Mary Jo's mother lets you have the back of their garage for a clubhouse.

One of the members has an uncle who is moving away to another city. He does not want to take with him his library of children's books. He gives the books to the club.

The day you get the books is very exciting. But it soon turns out that there just is not room for all of them in the clubhouse. You must build some more shelves. The members decide to divide the books up and each take home the same number until the shelves have been made.

You count the books. There are 237 of them. how many are there for each of the eleven club members?

You might start by guessing that there are enough for ten apiece. So you count out ten books for each of the eleven club members.

When you finish, you find that you have taken only 110 of the books. It is easy to see that taking these 110 from the original 237 leaves more than 110 left. So you know that there are at least enough for another ten each

After you choose these, there are still a few left. You count the leftovers and find that you have seventeen of them. You can see that this is enough for one more apiece with six left over.

27

But suppose that by now it has turned out that some of the books are very popular. Say that there are twenty-five of them which almost everyone would like to have. Can you think what you might do about them?

This is where it comes in handy, for once, to have a remainder.

Instead of giving the members one book apiece from that seventeen, you take back one book from each member. These eleven added to the seventeen make twenty-eight. And these twenty-eight books, which include the twenty-five extra-popular ones, can stay in the clubhouse and be the temporary club library.

Try the guessing-game way of dividing a box of 73 apples with six friends.

Try dividing the pencils in five twelve-pencil
bundles among nine people.

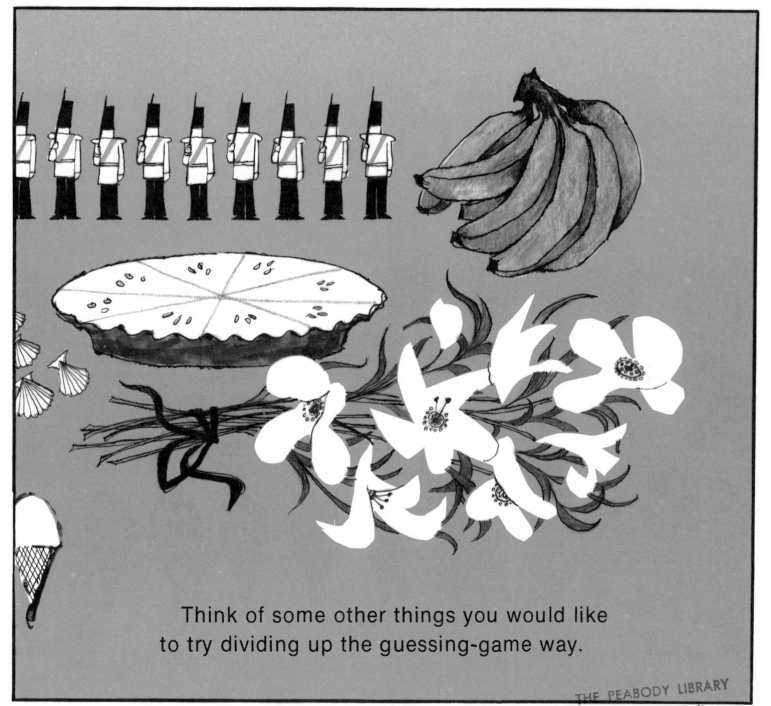

Think of some other things you would like
to try dividing up the guessing-game way.

ABOUT THE AUTHOR

This is Robert Froman's seventh book for the Young Math series. He says, "I always wanted to be a writer, and I had the luck to discover when I was a child that mathematics could be exciting, too. I like to pass that discovery along. The books I've done for the Young Math series are my favorite way of accomplishing this. What I've been trying to do in them is to make some of the basic ideas of mathematics both meaningful and intriguing to young readers."

ABOUT THE ILLUSTRATOR

A native New Yorker, Gioia Fiammenghi now lives in Paris, France, with her husband and children. She is a graduate of the Parsons School of Design, and has also studied at the Art Students League and in European schools.

Books illustrated by Gioia Fiammenghi have been chosen by *The New York Times*, the AIGA, and the Child Study Association in their lists of outstanding books for children.